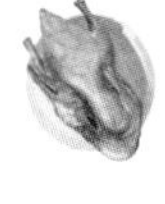

Chinese Cities
Nanjing Impressions

中国城市
南京印象

五洲传播出版社
China Intercontinental Press

The Story of Nanjing
故事
南京传奇

Visit to Historical Sites
访古
古迹往事

Traditional Urban Symbols
传统
城市符号

Humanity and Cultural Coordinates
人文
文化坐标

The Story of Nanjing

故事 南京传奇

Nanjing City Wall

An Ancient City Wall of the Ming Dynasty

南京明城墙 | 古老的城垣

Six hundred years ago, Zhu Yuanzhang, the founding emperor of the Ming Dynasty, made Nanjing the capital and oversaw the construction of the city walls, a project that took the dynasty more than 20 years to complete. Unlike the old ways of building square or rectangular city walls, Nanjing's city walls were built to align with the city's mountains and waters. The inner city wall, which was originally 35.3 km long, with 25.1 km of it completely preserved, is the longest city wall in China and the world.

The city wall embraces Nanjing and connects the past with the future. Its bricks and stones symbolize a condensed history of thousands of years.

600年前，明朝开国皇帝朱元璋定都南京，亲自监工，历时20余年修建城墙。明城墙营造一改以往方形或矩形旧制，而是依据南京的山脉、水系走向筑建，其中京城城墙长达35.3千米，现完整保存25.1千米，是中国规模最大的城墙，也是世界最长、规模最大的城垣。

城垣，构筑了家园。一头连着过去，一头接着未来，时光在砖石上流转，历史在这里凝结，回望千年。

More than 1,500 years ago, famous families of that era, used to live on Wuyi Lane and Zhuque Street near the Confucius Temple. This ancient city on the banks of the Qin Huai River has a spectacular history, and this flourishing river has inspired countless literati to write innumerable masterpieces and classics of literature that have left a mark on humanity.

Today, the river, which had witnessed the prosperity of the six dynasties, winds through the ancient city, and its babbling brooks sing songs of memories and times gone by.

早在1500多年前，夫子庙的乌衣巷、朱雀街就是名门望族聚居地。之后的千百年间，秦淮河两岸承载着金陵古城的万种风情，无数文人墨客在这条繁华的河流上，写下流芳百世的名篇。

如今，六朝金粉凝聚的十里秦淮，依然滋养着金陵古城。暮色四合、水汽氤氲，历史在画舫凌波中静静地荡漾。

Jiangnan Examination Hall

The Largest Imperial Examination Hall during the Ming and Qing Dynasties

江南贡院 | 明清最大的科举考场

"An unassuming farmer who succeeds in the imperial examination will gain both rank and wealth overnight." To succeed in the imperial exam was the dream of scholars in the ancient times. Jiangnan Examination Hall, the largest and most influential imperial examination hall in Chinese history, produced numerous scholars who ranked among the top three in the imperial exam during the hundreds of years that the examination was held since the Early Southern Song Dynasty till the Late Qing Dynasty when the imperial examination was finally abolished.

Now, the examination hall has become the Imperial Examination Museum of China, where China's only existing final imperial examination test paper written by a Number One Scholar is kept, allowing present and future generations to sneak a peek at the lives of scholars and literati of ancient China.

“朝为田舍郎，暮登天子堂”说出了多少旧时读书人的梦想。江南贡院，是中国历史上规模最大、影响最广的科举考场，南宋初建至晚清废除科举制的数百年间，从这里走出状元、榜眼、探花数不胜数。

如今的江南贡院已经成为中国科举博物馆，珍藏着全国仅存的殿试状元试卷，任凭后人想象着曾经的文人才子如烟尘一般的往事。

Xuanwu Lake

Jiangnan Royal Gardens and Lakes

玄武湖 | 江南皇家园林湖泊

"Indifferent to human love and hate, the willows peacefully dot the long bank." Known as the "Pearl of Jinling," Xuanwu Lake is the largest royal garden lake in south China. In spring, the lake, with its bluish waters rippling amid the shadows of the willows, looks like an ink and wash painting.

Around 1,500 years ago, during the Six Dynasties, the emperor trained his sailors here. Later, a royal garden was built by the riverside. Over the centuries, the lake was renamed several times as it went through moments of glory and decrepitude. The emperors Kangxi and Qianlong of the Qing Dynasty visited the lake and wrote poems in its praise. Later, regular construction of gardens and buildings led to the development of a unique area consisting of landscapes, cities, and buildings, which is popularly known as the "Five continents, one garden, one road."

"无情最是台城柳，依旧烟笼十里堤"。被誉为"金陵明珠"的玄武湖是中国仅存最大的江南皇家园林湖泊，春天的玄武湖上，环湖烟柳、碧波荡漾，如同一幅丹青水墨画。

早在1500多年的六朝时期，这里是皇帝操练水兵的场所，并被辟为皇家园林，后几度荒废，几度更名，又几度兴盛。清朝康熙、乾隆两位皇帝都曾经到访此处并吟诗作赋，之后，这里又不断修建园林建筑，最终造就了"五洲、一园、一路"，山水城林相融的绝色风景。

Jiming Temple

An Ancient Temple Witnessing Thousand Years of Vicissitudes

鸡鸣寺 | 千年古寺话沧桑

Four hundred eighty splendid temples still remain, of Southern Dynasties in the mist and rain.The Jiming Temple, known as "the first temple of the Southern Dynasties," and one of the oldest Buddhist and royal temples in Nanjing, remains a temple that sees good attendance even today, and is impressive to look at even after 1,700 years.

The ancient Jiming Temple was built during the Southern Dynasties (420 CE–589 CE) by Emperor Wu of the Liang Dynasty, as a royal temple. Emperor Wu became a monk and practiced Buddhism in the temple four times. Later, the temple was destroyed during a war. In the early Ming Dynasty, Zhu Yuanzhang, the founding emperor of the Ming Dynasty, had the temple rebuilt, restoring it to its former glory and reputation. Emperor Kangxi of the Qing Dynasty inscribed the words "Ancient Jiming Temple" at the site of the temple as its name, during his southern tour.

"南朝四百八十寺，多少楼台烟雨中"。有着"南朝第一寺"美誉的鸡鸣寺是南京最古老的梵刹和皇家寺庙之一，古寺历经1700多年风雨洗礼，至今依然香火鼎盛、气度依然。

古鸡鸣寺始建于南朝（420—589年），是梁武帝按皇家规制修建，他曾先后四次在寺内出家，之后古寺毁于战火。明朝初期，开国皇帝朱元璋复建古寺，以致其名声大振，重现往日荣光。清朝康熙帝南巡时为古寺题写"古鸡鸣寺"匾额。

Visit to Historical Sites

访古 古迹往事

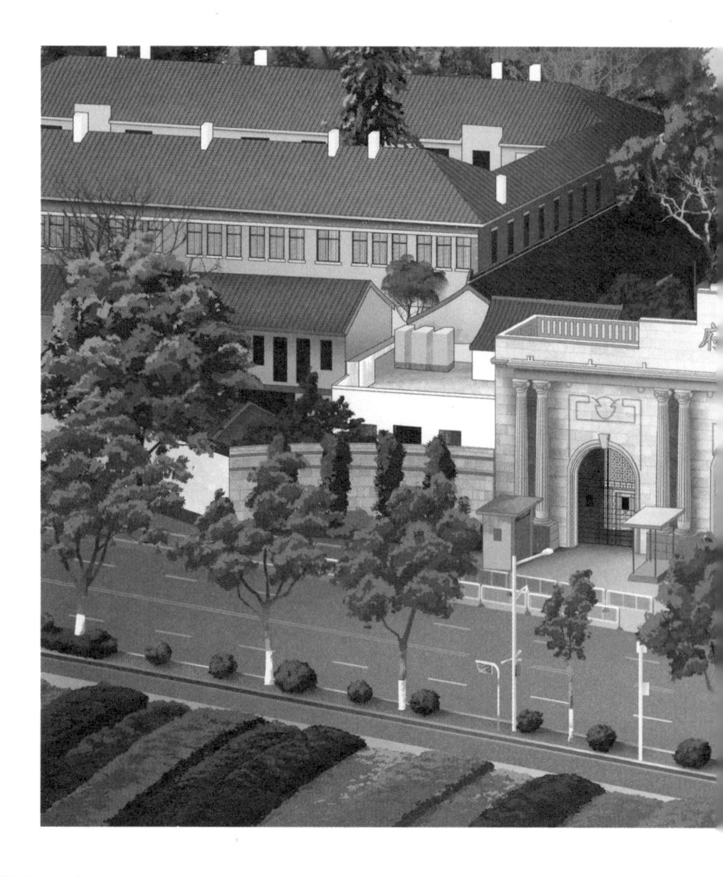

With a history of more than 600 years, Nanjing Presidential Palace is the largest and best-preserved architectural complex in modern China, and a model of the architecture of the Republic of China in Nanjing. Initially, it was a mansion for princes and marquises in the early Ming Dynasty. In the Qing Dynasty, it became the Jiangning Weaving Bureau and then office of the viceroy of Jiangnan Province and Jiangxi Province. Both Emperor Kangxi and Emperor Qianlong used it as a residence during their southern tour. The Palace has witnessed the turbulence of the past century of modern China.

People can visit the Presidential Palace, which is now a museum called the China Modern History Museum, and see the past unfold before their very eyes.

南京总统府已有600多年的历史，是中国近代建筑遗存中规模最大、保存最完整的建筑群，也是南京民国建筑的典范。最初为明初归德侯府和汉王府，清朝被辟为江宁织造署、两江总督署，清朝康熙、乾隆两位皇帝南巡时曾作为行宫，近代中国，这里又成为百年动荡历史的见证者。

这个古老的建筑群今天已经成为南京中国近代史遗址博物馆，向后人诉说着过去那些风起云涌的年代。

Sun Yat-sen Mausoleum

Reminiscences Amid Green Mountains and Clear Waters

中山陵 | 青山绿水间

Nestling in Zhongshan Mountain on the southern foot of Purple Mountain, is Sun Yat-sen Mausoleum, the tomb of Sun Yat-sen, the great forerunner of the democratic revolution. Viewed from the air, it looks like a "bell" lying sideways amid green mountains and clear waters. The solemn and serene mausoleum is one with the magnificent Zhongshan Mountain and is known as "the first mausoleum in the history of modern Chinese architecture."

Every detail and every building is emblematic of something. With the passage of time, the mausoleum, rather than fading away, has increasingly become a symbol of the intense collision of Chinese and Western architectural styles.

中山陵位于紫金山南麓钟山之中，是伟大的民主革命先行者孙中山先生的陵墓。从空中俯瞰，中山陵仿佛是一座平卧在青山绿水间的“自由钟”。宏伟的钟山与中山陵连成一体，庄严肃穆，被誉为“中国近代建筑史上的第一陵”。

在这里，每一个细节都有寓意，每一座建筑都有象征，时间并没有让它们褪去光泽，反而展现出中西建筑风格激烈碰撞的时代印记。

Xiaoling Mausoleum of the Ming Dynasty

The First Royal Mausoleum of Ming and Qing Dynasties

明孝陵 | 明清皇家第一陵

The Xiaoling Mausoleum of the Ming Dynasty, is the tomb of the Ming Emperor Zhu Yuanzhang and his empress. Surrounded by mountains and rivers, it was modeled on the prevalent practice of "building tombs in mountains" during the Tang and Song dynasties. To echo the three-courtyard pattern of the imperial palace featuring a "front court and rear resting place," the architectural pattern of the "front square and back circle" was created. Since Emperor Yongle moved his capital to Beijing, the tombs of Ming and Qing emperors were built the way the Xiaoling Mausoleum was built. That is why, it is known as the "first royal mausoleum of Ming and Qing dynasties".

In autumn when the skies are clear and tree leaves turn golden, the path to the mausoleum is covered with golden leaves, creating a serene yet charming atmosphere.

明孝陵是明朝皇帝朱元璋及其皇后的陵寝。山环水绕之中的明孝陵，承袭了唐宋两朝“依山为陵”的传统，又依照皇宫“前朝后寝”三进院落制，开创了“前方后圆”的建筑格局。明成祖迁都北京后，明清皇帝陵寝均按南京明孝陵的规制营建，所以明孝陵又被誉为“明清皇家第一陵”。

秋日的明孝陵，天高云淡、金风玉露、层林尽染，孝陵神道上铺满黄叶，别有一番韵味。

Qixia Mountain

A Sea of Colors

栖霞山 | 色彩之海

During the Southern Dynasties more than 1,500 years ago, Qixia Mountain got its name from the "Qixia Monastery," which was home to eminent monks of the time. Later, on, the Qixia Temple was built on the mountain and was visited by many emperors and nobles. Emperor Qianlong visited Qixia Mountain five times, and left an inscription in gold—"The First Jinling Mingxiu Mountain."

During the golden autumn, the red leaves dye the Qixia Mountain in golden hues. The beating of the evening drum and ringing of the morning bell at the Qixia Temple, the exquisite Zen-like ambiance of the dagoba—a dome-shaped shrine, and the magnificent Thousand-Buddha Grottoes, create a scene of serenity and tranquility, as depicted by Zhang Hong, a painter of the Ming Dynasty, in his work, the Qixia Mountain.

早在1500多年前的南朝时期，栖霞山就因高僧隐居的"栖霞精舍"而得名，后兴建栖霞寺，诸多皇帝和王公贵族登临，其中，五次驾临栖霞山的清朝乾隆皇帝还留下了"第一金陵明秀山"的金字。

深秋的栖霞山，漫山遍野的红叶层层叠叠、浓淡相宜。栖霞寺的暮鼓晨钟，舍利塔的精致禅意，千佛岩的恢弘气度，正如明朝画家张宏笔下的《栖霞山图》描绘的那般恬淡、幽远。

攝山栖霞寺
御筆

Meiling Palace, the former residence of the chairman of the Nationalist Government, was where Chiang Kai-shek and Soong Meiling lived for several years. In autumn, the leaves of the plane trees on either side of the Palace turn golden. When viewed from the air, they look like a jeweled necklace, with the Palace, a crystal emerald pendant in the center.

The luxurious and beautiful Meiling Palace perfectly matches the bright-colored plane trees, and is a historical and architectural masterpiece. Although its original owners are no more here today, the Palace, a building that reflects the aesthetics of the Republic of China and contains mellow historical and cultural connotations, continues to weave everyday stories into history.

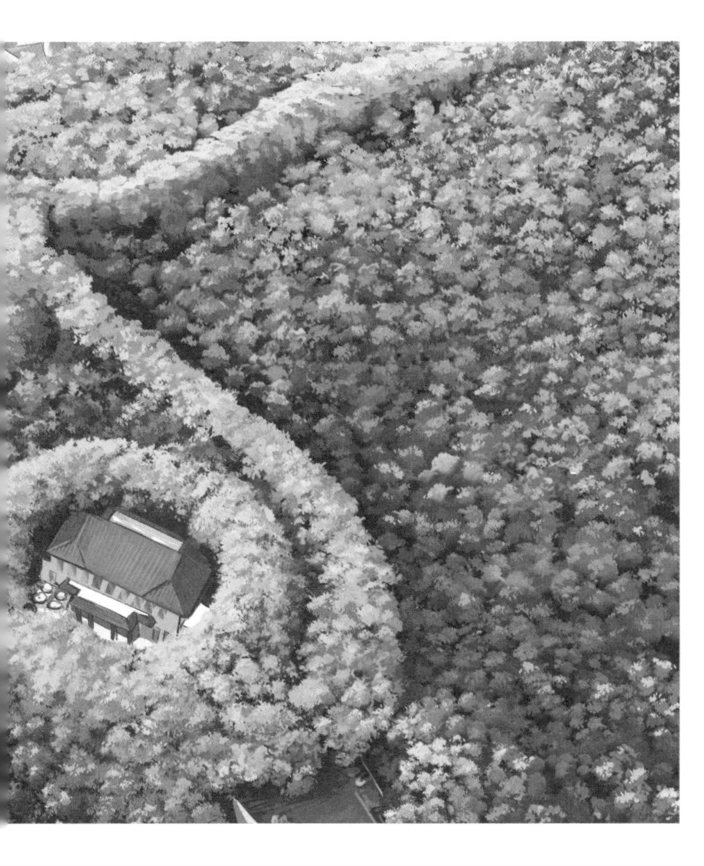

美龄宫是国民政府主席官邸旧址，蒋介石和宋美龄曾在此居住数年。金秋时节，美龄宫两侧的法国梧桐树叶变成了金黄色，从高空俯瞰，俨然一条宝石项链，而美龄宫就是那颗晶莹剔透的绿宝石挂坠。

奢侈漂亮的美龄宫，烂漫唯美的梧桐树，历史与建筑一脉相承。如今，这栋折射民国时期审美、包含醇厚的历史文化内涵的建筑，却是主人不复、历史绵延。

Traditional Urban Symbols

传统 城市符号

Qinhuai Lantern Festival

Sounds of Oars, Lights and Shadows

秦淮灯会 | 桨声灯影的记忆

"Tens of thousands of lamps light the city, making it shine as bright as the Milky Way." The Qinhuai Lantern Festival dates to the Southern Dynasties more than 1,500 years ago, with its heydays during the Ming Dynasty. The Lantern Festival is celebrated on the evening of January 15 of the lunar year, when "all citizens come out to the riverside to admire the lanterns,"—it is as though half of Nanjing is lit up in a glistening light of waves.

The Qinhuai Lantern Festival, a shining pearl of the intangible cultural heritage, carries people's sweet memories of the brilliant evenings of old Nanjing.

“汇数万火盏，若星河灿天衢”。秦淮灯会的历史最早可以追溯到1500多年前的南朝时期，明朝达到了鼎盛。每逢正月十五元宵节，夜色阑珊之时，“家家走桥，人人看灯”，大半个南京城都笼罩在波光水色的映衬中。

秦淮灯会成为非物质文化遗产中一颗璀璨的明珠，承载着人们对旧时南京的依稀记忆。

Yun Brocade

As Valuable as Gold

云锦 | 寸锦寸金

"Picturesque Jiangnan is famous for its brocade—created through exquisite craftsmanship and in a variety of patterns, such as peacock, silkworm, and dragon." After Wu Meicun, a scholar during the late Ming Dynasty, saw the Nanjing brocade, he wrote a song in its praise. Nanjing brocade dates to the Eastern Jin Dynasty (317 CE–420 CE). The brocade would be woven with exquisite and expensive materials, including pure gold thread, silver thread, copper wire, filament, spun silk, as well as peacock velvet and other birds and animals' feathers. As a result, it was extremely expensive and known as "each inch is worth an inch of gold." That is why during the Ming and Qing dynasties, the emperor's dragon robe was the most frequently ordered "personally tailored" brocade. With its layers of gold and the resplendent red and royal blues, the brocade added radiance and beauty to the person who adorned it and is indeed worthy of the title "An Oriental Treasure."

"江南好，机杼夺天工，孔雀妆花云锦烂，冰蚕吐凤雾绡空，新样小团龙。"这是明末文人吴梅村见到南京云锦后所写。南京云锦历史可以追溯到东晋时期，由于云锦在织造中用料不惜成本，使用纯金线、银线、铜线及长丝、绢丝，包括孔雀丝绒等各种鸟兽羽毛，造价极其昂贵，被誉为"寸锦寸金"。正因为此，明清时期，皇帝龙袍成为云锦最常接到的"私人订制"。云锦层层叠叠的金色、红色和宝蓝色交相辉映、光彩夺目，不愧为世界的"东方瑰宝"。

Baiju (ancient folk music of Nanjing)

Remembering the Sweet Old Days

白局 | 旧日的光阴

As the workers stepped on the complicated looms, they would hum graceful Jiangnan melodies, such as Baiju, to amuse themselves and break the dullness of the work. As a Nanjing-rooted folk art, Baiju, a melody that imbues elements of folk art, news of current affairs, food landscapes, and festival folk customs, brought amusement to Nanjing's residents with the use of the old Nanjing dialect. During the heydays of brocade culture, textile workers earned a lot. They sung such melodies at weddings or funerals free of charge, hence the name Baiju (literally meaning performing free of charge). In 2006, Baiju was included in Jiangsu's Intangible Cultural Heritage List.

踩着云锦的繁复机车，哼着婉约的江南小调，“白局”源于织锦房机工单调工作中的自娱自乐。作为土生土长的南京民间艺术，时事新闻、美食风景、节庆民俗，都是“白局”常见的题材，把南京人的乐子用老南京的方言说唱给南京人听。云锦鼎盛时期，纺织工人收入颇丰，因红白喜事“受请不受物，白唱不卖钱”，“白局”而得名。2006 年，白局入选江苏省非物质文化遗名录。

Velvet Flowers

An Intangible Heritage Borne out of Fingertips

绒花 | 指尖上的非遗

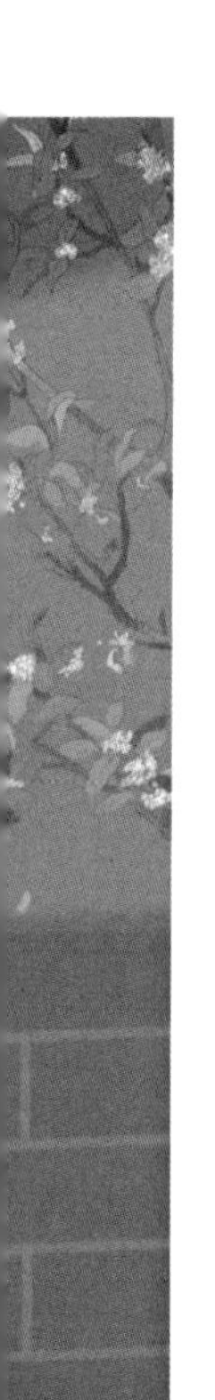

The word velvet flower, which sounds like rong hua in Chinese (two Chinese characters meaning wealth and status), was a traditional handicraft made of silk and copper wire. As early as the Tang Dynasty, velvet flowers were designated as royal offerings. During the Ming and Qing Dynasties, brocade flourished in Nanjing. Due to the strict standards for selection of materials, during the weaving process, a lot of silk scraps would be left, which were then made into velvet flowers with delicate shapes by the local artisans of that time, who hated to see anything wasted.

Empress Fucha of Emperor Qianlong liked "decorating herself with velvet flowers, rather than royal jade." Later, velvet flowers were used by ordinary folks, especially by brides who used them in their headdress. It is true that a bride who wears velvet flowers has an ethereal charm.

绒花，谐音“荣华”，是用蚕丝和铜丝制作的传统手工艺品，早在唐朝时，绒花便被定为皇室贡品。明清时期，南京云锦鼎盛，由于选材严苛，织造过程中会留下大量的蚕丝边角料，古人惜物，就做成了造型灵动的绒花。

清乾隆皇帝的富察皇后喜欢“以通草绒花为饰，不御珠翠”。后来，绒花流入民间，成为人见人羡的新娘头饰，粉黛云鬓，头戴绒花的新娘，显得格外妩媚。

Salted Duck

The Perfect Dish for All Seasons

盐水鸭 | 好卤知时节

If food is a city's soul, then Nanjing's soul must be duck.

A standard salted duck must have "white skin, red meat, and green bones," the meat must be tender, and the marinade must be old. Some old shops are famous for their century-old marinade. Salted duck during the Mid-Autumn Festival holiday is known as "osmanthus duck," because when osmanthus blossoms, the duck is mostly white and fat, and tastes especially delicious after being put in brine.

如果说一个城市的灵魂是美食赋予的，那么南京的灵魂一定是鸭子。

一只标准的盐水鸭，必须符合“皮白肉红骨头绿”三大要求，鸭肉一定要够嫩，卤汁一定要够老，有些老店之所以声名显赫，就是因为拥有了镇店之宝百年老卤。中秋时节的盐水鸭，有着“桂花鸭”的美称，因桂花绽放之时，鸭肉最为洁白肥美，卤后才能愈加唇齿生香。

Humanity and
Cultural Coordinates

人文 文化坐标

Librairie Avant-Garde (Mount Wutai Branch)

The Soul Is a Stranger on Earth

先锋书店 | 大地的异乡者

Librairie Avant-Garde (Mount Wutai Branch) has been rated by overseas media as "the most beautiful bookstore in the world." It is one of the most beautiful cultural landmarks in Nanjing. The slogan of the bookstore—"The Soul Is a Stranger on Earth" is a line from Austrian poet George Trakl's poem *Springtime of the Soul*, and contains an implied meaning that the human spirit is always looking for a hometown that is nowhere to be found.

On a beautiful afternoon, in this warm and quiet space, you can pick up a favorite book, sip a cup of aromatic coffee, or listen to a cultural talk. Whatever you do, you will feel yourself completely refreshed, as if all your troubles and concerns have been blown away.

南京先锋书店（五台山店）被海外媒体选为“全球最美书店”，也是南京最美的文化地标。奥地利诗人特拉克尔的诗句“大地上的异乡者”是先锋书店的店铭，寓意人的精神永远在寻觅一个无所在的故乡。

明媚的午后，徜徉在温暖安静的空间中，挑一本心爱的书，抑或捧上一杯氤氲的咖啡，听一场文化讲座，心底的阴霾，仿佛也会被拿出来彻底晾干。

Nanjing University
The Story of Youth

南京大学 | 青春的故事

"Ivy grows all over the Peking University Building. But is it a vine or the parting sorrow of a wanderer?"—a description of the Peking University Building by poet Yu Guangzhong in his poem, was built in 1917 and was the tallest building at that time. It remains one of Nanjing's landmarks. The predecessor of Nanjing University was the Sanjiang Normal School founded in 1902. Today, Nanjing University has four campuses—Xianlin Campus, Gulou Campus, Pukou Campus, and Suzhou Campus. The Gulou Campus has the largest number of old buildings. The grand auditorium and small auditorium, both century old and designed in a combination of Chinese and Western styles, had their exterior walls made of Ming Dynasty wall bricks—the inscriptions on the bricks reflect the imprint of time.

The same campus has conceived countless dreams. More than 100 years have passed, and many alumni of the university have grown old, but the young men and women in their prime are walking on the footsteps of those before them, as they continue to write the story of youth.

"常春藤攀满了北大楼，是藤呢？还是浪子的离愁"。诗人余光中吟诵的北大楼诞生于 1917 年，是当时最高大的建筑，至今仍然是南京城的地标。南京大学的前身，是创建于 1902 年的三江师范学堂。现在南京大学拥有仙林、鼓楼、浦口、苏州四个校区，鼓楼校区的老建筑最集中。大礼堂和小礼堂历经百年风华，设计风格中西合璧，外墙全部采用明代城墙砖，砖上的铭文蕴藏岁月印记。

同一个校园，无数个梦想，百余年时光流逝，曾经的南大学子已是耄耋老人，但风华正茂的少年，仍然走在前辈走过的道路上，青春的故事，永不散场。

Nanjing Museum
The Glory of the Ancient Capital of Six Dynasties

南京博物院 | 六朝古都的华彩

Founded in the 1930s, Nanjing Museum houses a history of Jiangnan from the Stone Age to the Ming and Qing dynasties. The gold and silver interlaced heavy copper kettle from the Warring States Period, the silver wisp jade dress from the Eastern Han Dynasty, and the underglaze red plum vase patterned with pine, bamboo and plum from the Ming Dynasty, are some of the treasures of the museum.

The most distinctive section is the Republic of China Hall, where rickshaws, old post offices, railway stations, barber shops, medicine shops, bookstores, and coffee shops of that era are in display... Visitors to the hall, take a journey back in time to a world of cheongsams and qipaos, of people relaxing on benches in old-styled cafes, listening to quaint music from old records—it is as though the old days of Nanjing a hundred years ago have come back to life...

南京博物院始创于20世纪30年代，展示了从石器时代直至明清江南的历史积淀。战国时期的错金银重络铜壶、东汉时期的银缕玉衣，明朝的釉里红岁寒三友纹梅瓶等，都是镇馆之宝。

南京博物馆里最具特色的当属民国馆，黄包车、老邮局、火车站、理发店、药铺、书店、咖啡厅……仿佛来客换上一身旗袍或长衫，坐在老式咖啡馆的长凳上，听一帧老唱片，就穿越回了百年前南京的旧时光之中。

Yihe Road

Century-old Western-style Buildings and Phoenix Trees

颐和路 | 百年洋楼和梧桐

In 1872, a French missionary planted Nanjing's first plane tree on Shigu Road. Since then, tens of thousands of plane trees have gradually become an integral part of the city with the passing of time, offering a backdrop for innumerable romantic legends. The plane trees on Yihe Road are the most beautiful ones in Nanjing. In the 1920s and 30s, more than 200 garden houses towered the road, earning the name "century-old Yihe, epitome of the world." Today, these old buildings, nestling under the shadows of the plane trees, are symbols of the vicissitudes of life as well as a life of luxury.

1872年，一位法国传教士在石鼓路种下了南京第一棵法国梧桐树，自此，数万棵梧桐，在淡去的光阴里渐渐融入这座城，也衍生了诸多浪漫的传说。颐和路的梧桐树，是整个南京最美的所在。“百年颐和，万国风华”，20世纪二三十年代，这里云集了200多座花园洋房。如今的这些建筑修旧如旧，掩映在亭亭如盖的梧桐树中，更显得沧桑而华贵。

Zhongshan Avenue
The Earliest Zhongshan Road

中山大道 | 中国首条中山路

It is said that "There are as many Zhongshan Roads as there are cities in China." Built in 1928, Zhongshan Avenue in Nanjing, changed the spatial structure of the ancient city. As one of the few modern urban roads in China at that time, it was reputed as the "Meridian of the Republic of China."

The old buildings of the Republic of China stand quietly on both sides of Zhongshan Road, not the least outshined by the nearby high-rise buildings. At the end of the avenue a bronze statue of Sun Yat-sen stands tall, gazing down at the bustling new street. Nanjing, a city where history and modernity mingle, is radiant with new vitality.

有人说："中国有多少城市，就有多少中山路。" 1928 年建成的南京中山大道，改变了这座古城的空间结构，是当时中国屈指可数的现代化城市干道之一，曾被赋予"民国子午线"的美称。

如今，中山路两旁保留下的民国老建筑，点缀在高楼林立之间，路尽头伫立着的孙中山先生铜像，凝视着车流如织、繁华喧闹的新街口。这座历史与现代交相辉映的城市，正在焕发着新的活力。

Walmart

图书在版编目（C I P）数据

南京印象 : 汉英对照 / 陈曦著 . -- 北京 : 五洲传播出版社 , 2022.7
（中国城市）
ISBN 978-7-5085-4819-7

Ⅰ . ①南… Ⅱ . ①陈… Ⅲ . ①本册 Ⅳ . ① TS951.5

中国版本图书馆 CIP 数据核字 (2022) 第 066785 号

中国城市：南京印象
Chinese Cities：Nanjing Impressions

出 版 人：关　宏
责任编辑：杨　雪
助理编辑：刘婷婷
插　　画：王建华　黄小升
文　　字：陈　曦
译　　者：陈跃鹏
设计策划：山谷有魚
装　　帧：张伯阳
出版发行：五洲传播出版社
地　　址：北京市海淀区北三环中路 31 号生产力大楼 B 座 6 层
邮　　编：100088
发行电话：010-82005927，010-82007837
网　　址：http://www.cicc.org.cn，http://www.thatsbooks.com
印　　刷：北京市房山腾龙印刷厂
版　　次：2022 年 7 月第 1 版第 1 次
I S B N：978-7-5085-4819-7
开　　本：889mm × 1194mm　1/32
印　　张：6
字　　数：20 千
定　　价：49.8 元